WITHDRAWN
BY
WILLIAMSBURG REGIONAL LIBRARY

MARINE CORPS FORCES SPECIAL OPERATIONS COMMAND

U.S. SPECIAL FORCES

MARINE CORPS FORCES SPECIAL OPERATIONS COMMAND

JIM WHITING

WILLIAMSBURG REGIONAL LIBRARY
7770 CROAKER ROAD
WILLIAMSBURG, VA 23188

CREATIVE EDUCATION

APR -- 2015

PUBLISHED BY Creative Education
P.O. Box 227, Mankato, Minnesota 56002
Creative Education is an imprint of The Creative Company
www.thecreativecompany.us

DESIGN AND PRODUCTION BY Christine Vanderbeek
ART DIRECTION BY Rita Marshall
PRINTED IN the United States of America

PHOTOGRAPHS BY
Alamy (Archive Image, Nate Derrick, GL Archive, Pictorial Press Ltd, Planetpix, US Air Force Photo, US Army Photo, US Marines Photo, US Navy Photo, Anatoly Vartanov), Corbis (George Steinmetz, Stocktrek Images), Getty Images (Getty Images, Jacob Silberberg), iStockphoto (spxChrome), Shutterstock (ALMAGAMI, gst), SuperStock (Ed Darack/Science Faction, PARAMOUNT PICTURES/FRENCH, KIMBERLY/Album/AI)

COPYRIGHT © 2015 CREATIVE EDUCATION
International copyright reserved in all countries.
No part of this book may be reproduced in any form without written permission from the publisher.

LIBRARY OF CONGRESS CATALOGING-IN-PUBLICATION DATA
Whiting, Jim.
Marine Corps Forces Special Operations Command / Jim Whiting.
p. cm. — (U.S. Special Forces)
Includes bibliographical references and index.
Summary: A chronological account of the American special forces unit known as Marine Special Ops, including key details about important figures, landmark missions, and controversies.

ISBN 978-1-60818-464-4
1. United States. Marine Special Operations Command—Juvenile literature. 2. Special forces (Military science)—United States—Juvenile literature. I. Title.

VE23.W44 2014
359.9'6—dc23 2013036220

CCSS: RI.5.1, 2, 3, 8; RH.6-8.4, 5, 6, 8

FIRST EDITION
9 8 7 6 5 4 3 2 1

TABLE OF CONTENTS

INTRODUCTION 9

TACTICS FROM TRIPOLI TO TODAY 11

TRAINING THE BEST OF THE BEST 19

TOOLS OF THE TRADE 27

NOTABLE MARSOC MISSIONS 35

★★★

GLOSSARY 45

SELECTED BIBLIOGRAPHY 47

WEBSITES 47

READ MORE 47

INDEX 48

U.S. Marines practice parachuting at California's Camp Pendleton and other locations.

FORCE FACTS Among the more challenging advanced training exercises are free-fall parachute operations, categorized as high altitude–low opening (HALO) or high altitude–high opening (HAHO).

INTRODUCTION

It's about 2:30 a.m. in a remote mountain village in Afghanistan. Suddenly, the silence is shattered by the sounds of revved-up dirt bike engines. Anyone who sticks his head out the door might be terrified by the sight of men—many of them bearded—dressed in combat gear and carrying up to 50 pounds (22.7 kg) of heavy-duty weaponry gunning their bikes down the village's main street. If that person is a peaceful villager, no problem. But if he is a member of the ***Taliban***, his life expectancy is probably measured in seconds.

These bikers are members of the Marine Special Operations Command, or MARSOC, an elite special force within the already elite Marine Corps, and they are turning the tables on the enemy. For years, the Taliban would appear out of nowhere on their dirt bikes, plant *improvised* explosive devices (IEDs) or fire at American troops, and flee to their hideaways—where they would prepare to strike again. In early 2012, MARSOC decided to "fight fire with fire." Several dozen "operators," as American special forces personnel are often called, learned to ride dirt bikes—some of them identical to the ones the Taliban used—over rough mountain trails, maintaining control during high-speed ascents and descents, avoiding obstacles such as large stones and tree roots, and roaring across swift-flowing creeks.

No one expected these "*Hells Angels* on *steroids*" to single-handedly defeat the Taliban. Rather, the team could win a series of small victories by using its enemy's tactics against them—another example of MARSOC's characteristic flexibility and creativity.

MARSOC operates in rough, unfamiliar terrain that is home only to terrorist camps.

TACTICS FROM TRIPOLI TO TODAY

The same spirit of turning the table on one's enemies has worked in America's favor before. In the early 1800s, pirates based on the coast of North Africa terrorized American merchant ships during voyages through the Mediterranean Sea. The pirates captured the ships and held the crews for ransom. In 1802, United States president Thomas Jefferson sent a fleet of American warships to fight the pirates. Unfortunately, the USS *Philadelphia*, one of the fleet's most powerful vessels, ran aground near the harbor of Tripoli (today the capital city of Libya) the following year and was captured. Navy lieutenant Stephen Decatur and several dozen volunteers—including a number of U.S. Marines—sailed a small captured pirate ship to the *Philadelphia* as it lay at anchor. They killed the guards and burned the ship. This daring action ensured that the *Philadelphia* couldn't be used against American forces.

A few months later in the same conflict, Marine Lieutenant Presley O'Bannon and eight *enlisted men* headed up a force of 500 *mercenaries* who sought to capture Tripoli. They had covered most of the distance to their objective—and defeated a much larger pirate force at the city of Derna—when a peace treaty ended the threat to American shipping. Part of the present-day Marine Corps hymn ("to the shores of Tripoli") pays tribute to that long-ago heroism.

The night of February 16, 1804, Decatur's team set fire to the captured Philadelphia.

FORCE FACTS Seven Medals of Honor and 134 Navy Crosses were awarded among the 8,000 men who served as Marine Raiders during World War II, a much higher percentage than conventional units.

The Marines' derring-do was also evident during the early days of American participation in World War II as part of the Marine Raider battalions. The Marine Raiders conducted missions behind Japanese lines, obtaining valuable intelligence about enemy defenses before invasions and launching lethal attacks. In summing up the reasons for the effectiveness of the 2nd Raider Battalion, Lieutenant Colonel Evans Carlson explained, "Most important, though, was the development of what we call the Gung Ho spirit; our ability to cooperate—work together.... This called for self-discipline and implicit belief in the doctrine of helping the other fellow. Followed through to its ultimate end it would mean that each while helping the other fellow would in turn be helped by him." The Gung Ho spirit continues to infuse the actions of individual Marines.

During the Vietnam War, where many battlefields had fluid borders without the established front lines of previous wars, the ability to go deep into enemy territory became even more vital. That ability led to the establishment of the 1st Amphibious *Reconnaissance* Company. Force Recon, as it became known, predated the Navy SEALs and the Army's Long Range Reconnaissance Patrols (LRRP). Force Recon personnel became adept at air and seaborne *infiltrations* as well as land tactics. Initially, Force Recon personnel tried to avoid direct conflict with the enemy. However, they increasingly became involved in firefights and inflicted nearly six times as many casualties during the conflict as regular Marine units. Force Recon remains in existence and has served with distinction most recently in Iraq and Afghanistan.

As the years passed, the nature of warfare changed further. Small groups of terrorists were able to inflict considerable physical and psychological damage on their targets. Traditional military tactics proved ineffectual in the struggle to control such threats, and the number of terrorist acts increased steadily. By the mid-1980s, the need for a coordinated command to oversee

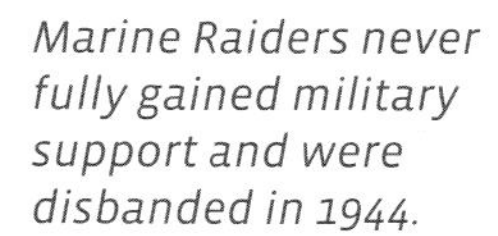

Marine Raiders never fully gained military support and were disbanded in 1944.

FORCE FACTS Force Recon continues to exist within the Marine Corps, and its primary field of responsibility is gathering intelligence for Marine expeditionary and *amphibious operations*.

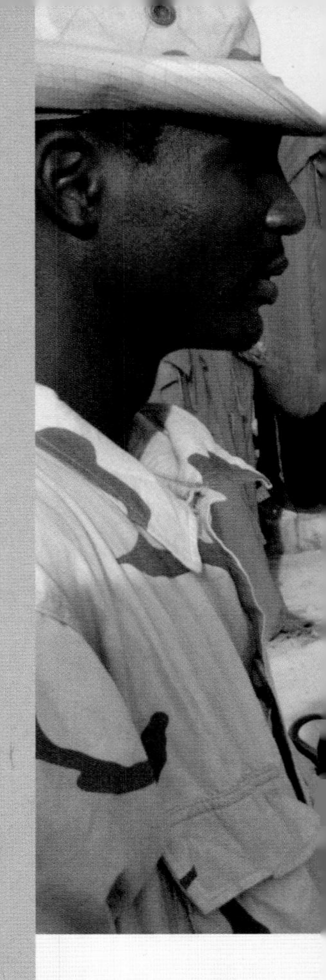

all special operations forces (SOFs) became apparent. That led to the formation of the U.S. Special Operations Command (SOCOM), a separate unit within the armed forces with its own command structure. At first, Marine Corps leaders had several objections to joining SOCOM, the first of which was that every Marine was capable of performing special operations on his own. Secondly, the Corps did not want to lose control over any units currently under its scope, such as Force Recon.

But as terrorism continued to rise—especially after the 9/11 attacks in New York and Washington, D.C., that killed thousands of civilians—all military branches faced facts that the number of missions exceeded the supply of special forces personnel. Because it was the only military service that wasn't yet part of SOCOM, the Marine Corps was the likeliest place to turn for additional boots on the ground. Secretary of defense Donald Rumsfeld brokered an agreement between SOCOM and the Marines that led to the formation of MSSOCOM Detachment 1, or Det 1. This unit was composed primarily of Force Recon Marines and served alongside Navy SEALs in Iraq. Its performance indicated that a special force of Marines could become a useful element of SOCOM. So at a press conference late in 2005, Rumsfeld announced to reporters, "In this complex and unconventional conflict, we are constantly looking for ways to strengthen our armed forces. One of the results of these studies is that I've just approved the creation of a Marine Corps component in the U.S. Special Operations Command." Known as MARSOC, the new unit became operational early in 2006, with an eventual authorized strength of 2,500 personnel.

MARSOC didn't waste any time in demon-

On September 11, 2001, terrorists flew two planes into New York's World Trade Center.

strating its value. Within a few months of its establishment—even before significant specialized training began—operators were dispatched to hot spots in Africa and South America on several missions to work with local troops under attack. Perhaps the most notable mission took place in the African country of Chad. Islamist rebels based in neighboring Sudan had been swarming across the border attempting to overthrow the existing government and possibly establish a base for terrorist operations. The rebels nearly reached the Chadian capital of N'Djamena before they were pushed back by two battalions that had been trained by MARSOC operatives. Speaking about how the operators had only a few weeks to work with the Chadians, then MARSOC commander Major General Dennis Hejlik said, "To bring that force up to that level was absolutely amazing."

U.S. operatives train foreign troops worldwide in local counterterrorism techniques.

By then, the basic building blocks of MARSOC had been established. Each Marine Special Operations Team (MSOT) consists of 14 men known as critical skills operators (CSOs). Four MSOTs form a company, and four companies comprise a Marine Special Operations Battalion (MSOB). Each MSOT includes a captain who commands the team, a team chief (master sergeant), an operations chief (gunnery sergeant), and a radioman. It also has two 5-man fire teams (a staff sergeant and three other sergeants and/or corporals—one of whom is armed with a *squad automatic weapon*—plus a navy corpsman). Currently, there are three MSOBs in the U.S. Two are headquartered at Camp Lejeune, North Carolina, and the third is in San Diego, California. Collectively, they make up the Marine Special Operations Regiment, which serves as the heart of MARSOC.

The CSOs, nicknamed "doorkickers," are aided by the Marine Special Operations Support Group (MSOSG). One element of MSOSG is a logistics company that ensures the operators have plenty of ammunition, food, and any other mission-critical supplies. A support company provides communications and joint terminal attack controllers (JTACs), who are trained to call in air strikes or artillery fire in support of MARSOC missions. "At its core, a JTAC is an individual who is qualified and certified to direct the actions of combat aircraft engaged in close air support and other offensive air operations," according to MARSOC Public Affairs. "On a Marine Special Operations Team, however, JTACs are not only experts in air to surface fires, they are also subject matter experts in every function of aviation support." An intelligence battalion, which offers a wide range of information about mission objectives in real time, and the Marine Special Operations School—which selects and trains CSOs—round out MARSOC's organizational structure.

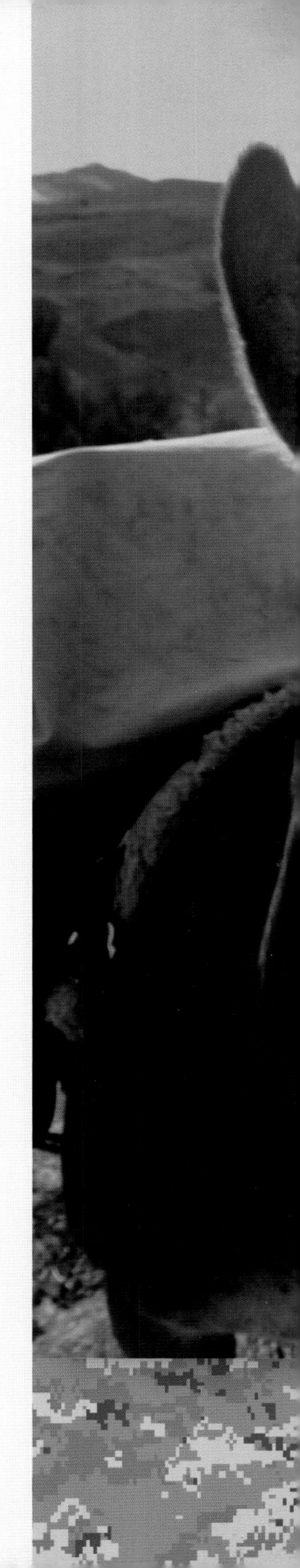

Animals can transport supplies and people through areas unreachable by motor vehicles.

FORCE FACTS Because CSOs are likely to be deployed to relatively primitive societies, learning to work with pack mules is an important part of their training.

TRAINING THE BEST OF THE BEST

MARSOC NEEDS TO ATTRACT THE RIGHT KIND OF PERSONnel and provide them with the best possible training so they will be "comfortable in the uncertainty of combat." It draws its recruits—all of whom are men—entirely from within the Marine Corps. As a result, potential operators begin their training with one built-in advantage: Every Marine is a qualified rifleman, which means that potential MARSOC operators already know how to shoot—and shoot well. That's a solid foundation for men who will eventually be called upon to utilize those shooting skills at a moment's notice in potentially life-threatening situations.

But marksmanship is just one quality of a successful operator. As General Hejlik noted soon after the formation of MARSOC, the outfit "does not directly recruit; it is dependent on the USMC [United States Marine Corps] and USN [United States Navy] for mature personnel well into their first 5 to 10 years of service.... The attributes MARSOC is looking for are those that enable the candidate to accept the greater responsibility of conducting special operations missions independent from large formations to a unique set of conditions and standards." For many MARSOC recruits, the motivation to join is simple. They want to push themselves even harder than their conventional training provides. They seek to test the limits of their endurance and to take the skills they already possess to an even higher level. "I joined MARSOC for the same reason that I became a Marine—intense,

Drill instructors instill the values of the Marine Corps, pushing recruits to their limits.

FORCE FACTS Operators continually practice FRIES (fast rope insertion/extraction system), using wool ropes from a helicopter to get in and get out of a combat zone in a few seconds.

determined people who aren't afraid to win," explained a corporal assigned to the 3rd MSOB. "MARSOC was the opportunity to work in a small group of capable individuals who are allowed to think and act with skilled creativity."

MARSOC leaders continually emphasize that they will accept only the most highly qualified candidates, so potential recruits undergo an exhaustive process before being accepted for training. It begins with a thorough screening of candidates' personnel records to ascertain that they are physically, psychologically, and medically qualified for the rigorous standards that MARSOC demands. The next step is assessment and selection (A&S), which is conducted several times annually. For nearly three weeks at a time, every prospective operator is required to pass a variety of physical and mental tests, under the eyes of instructors who are keenly aware of what operators require for success in the field. A&S is so intense that it requires a three-week introductory session known as the Assessment and Selection Preparation and Orientation Course.

What lies ahead for those who pass A&S is the individual training course (ITC). For the next seven months, candidates replicate a variety of situations they are likely to face in the field. According to the SOCOM handbook, the ultimate objective of training is to produce CSOs who "are complex problem solvers able to operate across the entire spectrum of Special Operations in small teams under ambiguous, sometimes austere, environments while maintaining a high level of mental flexibility and physical endurance." To achieve this goal, there's plenty of hands-on training and one-on-one interaction with instructors, with a goal of producing about 165 new operators every year. ITC takes a building-block approach, with four phases of ever-increasing stress levels indicative of the strains the men are likely to face during their deployments. In addition, language training is ongoing during the entirety of the program.

Marines train to execute land, sea, and air operations with perfect coordination.

FORCE FACTS To go ashore from the sea into enemy territory, operators practice paddling rigid-hulled rubber boats called Zodiacs with one man calling out the stroke cadence.

The first part of ITC lasts for 10 weeks and focuses on the basic skill sets all MARSOC operators must develop. It builds on the trainee's earlier Marine Corps training in physical fitness, swimming, and hand-to-hand combat. It also introduces field skills such as land navigation, mission planning, communications, and fire support. One of the most important programs during this phase is the Survival, Evasion, Resistance, and Escape (SERE) course. It focuses on survival and evasion skills in hostile environments; if operators are captured, techniques of resisting interrogation and exploitation while simultaneously planning an escape are also covered. Another vital skill is Tactical Combat Casualty Care (TACC), which instructs trainees in *trauma* care and techniques that can be used to save the lives of wounded comrades under battlefield conditions.

The second phase of ITC concentrates on small-unit tactics, which involves the combat movements of platoons and smaller units, and lasts for eight weeks. It includes training in amphibious operations such as small boat and scout swimmer techniques, combat photography, demolitions, and intelligence gathering and reporting. During this period, trainees undergo two exercises: "Operation Raider Spirit," which lasts for two weeks and simulates high-stress patrol and combat situations, and "Operation Stingray Fury," which concentrates on reconnaissance in a variety of urban and rural environments.

In simulated boat assaults, Marines approach ships in combat rubber raiding crafts.

ITC's third phase lasts five weeks and emphasizes close quarters combat (CQC). The trainees learn the techniques and tactics of engaging the enemy at close range, using a variety of weapons. At the end of this phase, the trainees participate in "Operation Guile Strike," a series of fast-moving raids that—like Phase 2—are staged in different places.

Marines gain versatility by learning how to operate in water for amphibious insertions.

The seven-week final phase concentrates on "irregular warfare." This involves conflicts with enemy combatants who aren't regular military forces that belong to a particular country. These combatants are often trying to overthrow a particular government and seize power for themselves.

MARSOC constantly seeks ways of improving both the quality and the safety of operator training. One prominent example of achieving these goals is the 4,000-square-foot (372 sq m) live-fire shoot house that opened in February 2012 at Camp Lejeune. The facility features smoke extractors, overhead catwalks for instructors to observe the action, and cameras that allow quick reviews of the situations the men have just encountered, thereby providing immediate valuable feedback. From a safety standpoint, the shoot house is a huge upgrade from its predecessor. Bullet-absorbing walls virtually eliminate the possibility of ricochets and flying bullet fragments, which allows operators to focus entirely on the tasks at hand.

One of those tasks is entering a room and quickly being able to distinguish between armed and unarmed people, and then quickly eliminating the dangerous ones. Although MARSOC operators have years of firearms practice, this type of environment presents a unique set of challenges. As training officer Captain Matthew Deffenbaugh notes, "Even for infantry guys, [CQC] is new. The shooting is in closer proximity than they've ever experienced. It's a dynamic environment."

Other new facilities which aid the trainees in developing needed skills include a four-story building that permits additional practice in urban combat, an outdoor pistol and rifle range, and a tower that enables a variety of activities such as rappelling and *fast-roping* from a helicopter.

The ITC's "final exam" is known as the "Derna Bridge" exercise. Lasting for up to three weeks, the exercise takes place over hundreds of square miles—many of them forested—in North and South Carolina. It demands the full application of all the skills the potential operator has learned and developed during the previous seven months. Operators link up with forces from a "partner nation," as many of their missions involve close cooperation with local troops. It's common for residents of the area in which the exercise takes place to volunteer and provide realism, enabling the Marines the opportunity to encounter dilemmas they may face during a deployment. After "graduating," the trainee becomes a critical skills operator (CSO) and is assigned to one of the three MARSOC battalions.

Being a CSO involves far more than learning how to survive and thrive on the battlefield. As the official MARSOC website observes, "CSOs can be called on to be: negotiators, advisors, teachers, problem solvers, and warriors. MARSOC operators must execute sound judgment at the tactical, operational, and strategic level of war, simultaneously." To achieve that end, MARSOC operators never stop training. They continually reinforce the skills they acquired during ITC with the MSOB to which they are assigned. In addition, they are encouraged to attend advanced schools in specific areas such as reconnaissance, sniping, parachute jumping, combatant diving, and intensive language training.

★★★

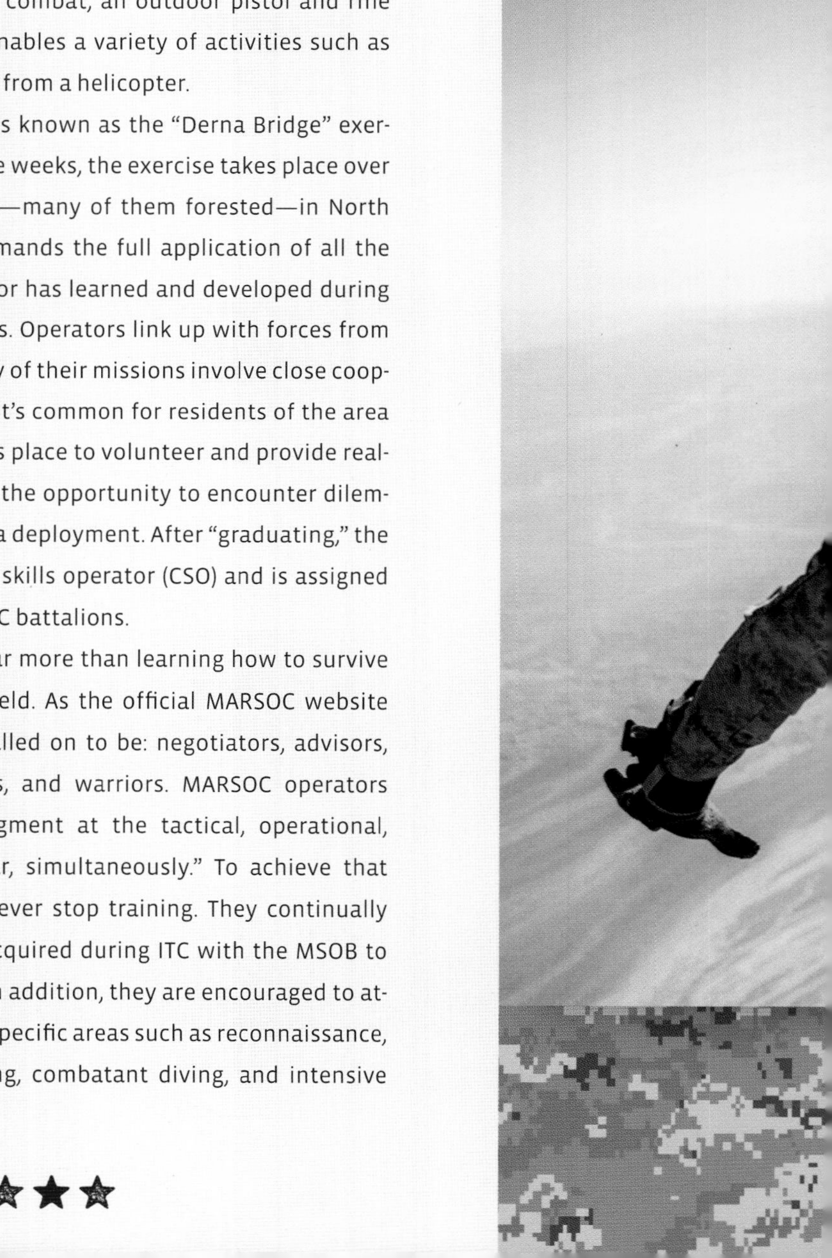

In HAHO parachuting operations, Marines open their chutes soon after jumping.

FORCE FACTS MARSOC operators agree to a five-year term of service to give themselves time to increase their skills and assume greater responsibility as they face new challenges.

TOOLS OF THE TRADE

Once a new operator has been assigned to his respective battalion, he can look forward to three primary types of missions. The most common—and by far the best known—is direct action. This involves short-term offensive actions intended to eliminate or severely damage enemy positions while inflicting as many casualties as possible. As MARSOC's initial mission in 2006 demonstrated, foreign internal defense—offering training and other assistance to foreign governments so they can assume responsibility for their own security—is another vital aspect, and it became especially important in Afghanistan. A third is special reconnaissance, conducting secretive operations to ascertain enemy strengths and the disposition of their forces.

Clearly, then, not everything MARSOC does involves "door-kicking." As American troops began accelerating the pace of withdrawal from Afghanistan in 2011, MARSOC added a new dimension to its missions in the form of village stability operations (VSOs), which involve living in Afghan villages and helping the people in such tasks as constructing schools and improving their economic situation. As Lieutenant Colonel Brian Petit, who commanded a Special Operations Task Force, explains, "We establish stability in the villages first, then connect village governance to the districts and the provinces. Investing in Afghanistan's villages is analytically rigorous, socially tiring, and highly

Sometimes protecting a place involves using camouflaged sharpshooters.

FORCE FACTS The M4A1 can be fitted with a quick-attach sound suppresser to greatly reduce the firing noise and the muzzle flash.

dangerous. Yet the rewards are worth the risk, for in combating Afghanistan's rural *insurgency*, we cannot 'win' without support from the villages."

While the establishment of new missions such as VSOs continually expands MARSOC's horizons, one thing remains constant: the need for a variety of dependable, effective weapons. The basic weapon of MARSOC operators is the Colt M4A1 carbine, designed to fit the needs of Special Forces. Similar to the standard M16A2 assault rifle carried by conventional troops, it has a shorter barrel—14.5 inches (37 cm)—which reduces weight and increases the operator's speed of action without diminishing its capability for applying lethal force in CQC. The carbine can be set for either semiautomatic or fully automatic firing. The weapon is designed to take numerous modifications, such as the addition of an enhanced power scope and sighting mechanism. A grenade launcher can be fastened directly beneath the barrel, firing a dizzying array of 40-millimeter projectiles that add considerably to the M4A1's destructive force. Some of these projectiles can punch through light armored vehicles, while others fragment when they are fired and slice through anything directly in front of the operator. The men also employ a variety of night-vision devices.

The carbines are complemented by the M1911A1 .45-caliber pistol, which—especially in CQC—can become the weapon of choice. While in many cases, new weapons constantly replace older ones, this is not necessarily true of the M1911A1. Robert Coates, the commander of Det 1, said, "The 1911 was the design given by God to us through John M. Browning that represents the epitome [perfect example] of what a killing tool needs to be. It was true in 1911 and is true now." A newer version, the M1911A1 Rail Gun, can accommodate lights, lasers, and infrared devices that detect body heat. It also reduces recoil and adds safety features.

While training Afghan forces in 2013, Marines used machine guns such as the M240B.

FORCE FACTS Another important M4A1 accessory is the combat sling, which secures the carbine across the body while allowing the operator to immediately return fire if attacked.

Another vital element in the MARSOC arsenal is the M249 special purpose weapon, a light machine gun that packs a lot of punch even though it weighs just over four pounds (1.8 kg). It is especially important to the individual fire teams within an MSOT because of the amount of ammunition it can direct against an enemy. As General James Conway notes, "The M249 allows Marines to establish fire superiority in a firefight, forcing attackers to take cover. I'm going to be hard-pressed to get fire superiority over [the enemy], to keep his head down instead of him keeping mine down, because that two-hundred-round magazine just keeps on giving."

Other important weapons include the M240G medium machine gun, which weighs 24.2 pounds (11 kg) and can fire between 750 and 950 rounds a minute with deadly accuracy at targets as far away as 1.1 miles (1.8 km) when it is supported by the built-in tripod. Many snipers like the M40A3 rifle, which was designed for accuracy and ruggedness. Starting in 2009, the M40A5 began replacing its predecessor. It has the capacity for attaching a sound suppressor, keeping potential targets unsure of where the lethal attacks are coming from. Another tried-and-true Browning creation that dates back to the 1920s, the "Ma Deuce" M2 .50-caliber heavy machine gun fires up to 450 rounds per minute at distances of up to 3 miles (4.8 km). It is capable of ripping apart vehicles and buildings.

The M4A1 rifle adheres to the well-known Marine saying, "Every Marine a Rifleman."

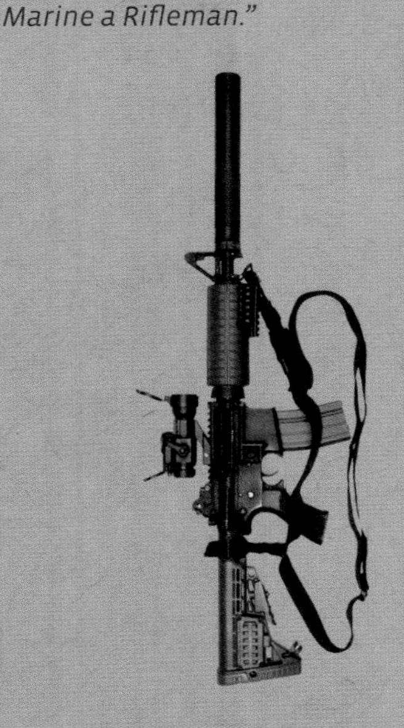

Having all this firepower—and usually quite a bit more—ensures that MARSOC operators enjoy a substantial advantage over their opponents. However, unlike other special forces such as the Green Berets and Navy SEALs, which have received extensive media coverage, MARSOC is too new and too secretive to be portrayed in the same manner. But its direct ancestor, Force Recon, has been an essential part of several films that feature big-name actors. *Heartbreak Ridge*, released in 1986, stars Clint Eastwood as a grizzled Marine sergeant during the invasion of

the Caribbean island nation of Grenada in 1983. Although movie critics liked it, many Marines felt the movie misrepresented their training and overemphasized the negative treatment that Eastwood's commanding officer gave him.

The movie Heartbreak Ridge *takes its title from a real-life Korean War battle in 1951.*

In *The Rock*, released a decade later, an angry ex-Marine brigadier general—who feels his nation has abandoned him and his men—leads a rogue group of Force Recon Marines who take over the former federal prison on Alcatraz Island and seize 81 hostages. They threaten to launch rockets laden with nerve gas into San Francisco. While the men's fighting skills were authentic, the film's premise seemed highly unlikely. One film critic said, "The movie deteriorates into a long commercial for the home-game version of itself," while another labeled it "[a] violence-intoxicated, far-fetched, morally corrupt drama."

In 2007's *Shooter*, actor Mark Wahlberg starred as a retired Force Recon sniper. He is framed for the assassination of an Ethiopian clergyman, a killing that becomes part of a larger conspiracy within the U.S. government. The film features a number of weapons used by Force Recon and MARSOC operators. More importantly, former Force Recon sniper Patrick Garrity worked with the filmmakers to ensure that Wahlberg's technical skills would replicate those of a true MARSOC sniper.

According to the film's production notes, "Garrity designed training as intensive as anything undertaken before by an actor preparing for such a role. He started by giving Wahlberg a basic history lesson about Marine scout snipers—who have made a difference in many American battles by providing precision fire on selected targets. Garrity then quickly moved into the nitty-gritty, teaching Wahlberg the use of different shooting positions, how to manipulate the *prismatic scope* for extreme long-range shooting, how to judge the effects of the wind on a shot and all the many vital factors that go into producing keen marksmanship."

Garrity even taught the actor the art of rapid-bolt manipulation, which involves firing as many shots as possible before the targets realize that they are being attacked. "That's one thing that Marine scout snipers have to do," he said. "They take their shot, rapidly manipulate the bolt and get right back on target.... it's almost like a dance step or art. You can really tell when somebody knows what they're doing with how they get set up for their shot."

Wahlberg learned still other skills, such as the importance of breathing to increase blood flow and help the sniper to relax before taking the shot. He often wore a ghillie suit, which is camouflaged clothing meant to mimic natural vegetation and allow the shooter to avoid detection even from observers standing a few feet away. Perhaps his sternest test came when, as he noted, "I had on 130 pounds [59 kg] of equipment, was carrying this huge sniper rifle, and had to walk out on this little, thin beam a couple of hundred feet in the air."

As the War on Terror continues and MARSOC keeps up its stellar work, it's likely that it will gain a higher profile in public consciousness. At that point, operators may become featured in films, books, video games, and other forms of media in much the same way as other special forces.

Shooter *received mixed reviews, as some praised the action and others cited plot holes.*

FORCE FACTS One of the key differences between the operations of MARSOC and traditional Marines is that MARSOC operators deploy independently to their operational area to plan and execute missions.

NOTABLE MARSOC MISSIONS

WHILE MARSOC'S ACCOMPLISHMENTS ARE WELL-KNOWN in the special forces community, they don't often gain publicity among the general population because so many missions are classified. Another factor is that the nature of MARSOC combat, especially in Afghanistan, doesn't lend itself to the type of large-scale operations that are the most likely to receive widespread media attention.

Perhaps MARSOC's best-known operation came in the town of Shewan in the Farah Province of western Afghanistan in 2009. Shewan had long been a Taliban stronghold, and more than 400 of its fighters in the immediate vicinity had already killed a number of troops from the coalition forces. A former Taliban member bragged to a MARSOC operator, "Many have tried to take Shewan and failed.... The Soviets tried to take Shewan. The coalition forces have already tried to take Shewan and failed. You will fail if you try to take Shewan."

One hundred twenty-five MARSOC operators and several hundred Afghan troops were given the task of proving him wrong. The operation—codenamed Red Thunder—began on September 26, 2009, with a variety of aircraft providing close air support for the ground forces. The fighting was so intense that the aircraft eventually ran out of ammunition. The aerial assault helped pave the way for slow but steady progress, with troops in many cases going house by house to clear the enemy. The operation came to a successful conclusion on October 1, with 65 Taliban dead and the rest having fled. One operator and three Afghan troops were killed. Significantly, not a single civilian per-

Afghan commandos (pictured) worked closely with Marines deployed to the country.

FORCE FACTS After Operation Red Thunder, village elders entered a white mosque for the first time in years and began planning how to rebuild their village.

ished in spite of the countless rounds of ammunition that both sides expended.

Testimony to the operation's effectiveness came from then-International Security Assistance Force Commander (ISAF) General Stanley McChrystal, who said, "The last time I was here in Farah, I heard firsthand about how bad Shewan was getting. This time, I got to actually go there and see the recent progress for myself, thanks to the noteworthy clearing operation there."

A year earlier, two MARSOC teams and a platoon of Afghan National Army commandos were dispatched to take down a Taliban warlord who had set up a stockpile of weapons in a remote canyon north of the city of Herat. Their objective was in a *wadi* so narrow that their Ground Mobility Vehicles (GMVs) would have been hard-pressed to turn around. As they approached the target, they encountered two empty vehicles blocking their way, and steep cliffs flanked either side. There was no way around the obstacle. Because there was a chance that the vehicles were booby-trapped, the operators decided against pushing them out of the way.

Moments later, a hail of machine gun bullets from hidden shooters wounded several men and forced everyone to seek shelter. As the firefight continued, one operator died and others suffered serious wounds. Staff Sergeant John S. Mosser pulled one badly wounded operator to safety under heavy fire, and then braved additional fire by standing in the open so that his *Global Positioning System* (GPS) unit could get an accurate fix on their position. He called in the *coordinates* on his radio, allowing support aircraft to bomb the position and enabling the unit to withdraw safely. Mosser received the Navy Cross, the nation's second-highest military decoration, for his heroism.

Much more typical of MARSOC, though, are small-group operations, in which the operators conduct patrols from *forward operating bases* (FOBs). They provide *humanitarian aid* to vil-

MARSOC provided security for Afghan law enforcement during construction projects.

FORCE FACTS Lt. Col. Evans Carlson, commander of the 2nd Raider Battalion, coined the phrase "Gung Ho!," the Corps' unofficial motto, from a Chinese expression meaning "work together."

lagers and seek to keep those villagers safe from the Taliban. The operators are constantly on the go, with breaks between missions lasting from 12 hours and up to only 2 or 3 days. As one anonymous operator said, "We don't like being on the FOB."

During one grueling FOB mission, which lasted for four days, MARSOC operators were attacked four times with a variety of weaponry: machine guns, mortars, and rocket-propelled grenades (RPGs). Despite the ferocious assaults, the group suffered no casualties. Afterward, the men went out again on a three-day mission to secure several nearby villages. The first day passed without incident. After an exhaustive search of one village, the men determined that no Taliban were present. They distributed the humanitarian aid they had brought with them.

The following day, the group came under attack as they entered another village. They quickly assumed defensive positions and engaged the enemy in a battle that lasted for four hours. Despite several close calls—an RPG missed one man by less than two feet (61 cm)—the operators killed some of their attackers and drove off the rest. After treating the villagers who had been injured during the fighting, the unit returned to its FOB for another brief respite before heading out again ... and again ... and again.

Marines on a forward operating base often secure Afghan villages from terrorists.

While this unit was fortunate to have conducted its missions without casualties, every engagement carries the risk of death. In June 2011, Corporal William Woitowicz and his team, along with a unit of Afghan Local Police, were assigned to attack the Taliban-controlled village of Panerak in Badghis Province. Within minutes of being helicoptered in, the men came under heavy fire. They were pinned down in a fairly exposed position. A rock wall across a dry riverbed would provide shelter

Village walls provide cover from not only firefights but also desert sandstorms.

and the opportunity to regroup, but two Afghani police had already been gunned down as they tried to reach the wall.

Woitowicz, who had shocked his family by joining the Marines immediately after his high school graduation four years earlier, didn't hesitate. He exposed himself to enemy fire as he laid down covering fire, allowing the rest of the men to make the crossing. At that point, Woitowicz tried to sprint to safety, but he was shot and killed as he tried to clamber over the wall. He was posthumously promoted to sergeant and awarded the Silver Star, the nation's third-highest military decoration.

A four-week period during July and August 2012 was especially lethal, as eight operators died. Three were killed in "routine" operations such as the ones described previously. The other five died in what are known as "green-on-blue" attacks, in which Afghan troops being trained to defend villages suddenly turn on the men who are working with them. The deaths brought the toll to 23 in the 6 years since MARSOC was established, with nearly 150 more operators being wounded.

Commenting on the upsurge in violence against his men during this period, MARSOC commander Major General Paul Lefebvre noted that his operators had had a significant effect in reducing violence. Yet he also noted the cost: “When you lose a MARSOC operator in small teams, you’re as close as close can be,” the general said. “Some of our guys have seven, eight, nine combat tours. They’re very, very experienced in terms of what they do. When you lose a guy like that, that’s hard. But you know what? There are 10 other men who will step up and do his job if we ask him to. They’re amazingly resilient in terms of what they’re doing.”

MARSOC made a rare appearance in the headlines in early December 2012. According to media reports, Sergeant William Soutra Jr. received the Navy Cross for his actions in Helmand Province. Three other members of Soutra’s team received the Silver Star. After Soutra and his team were ambushed, and the team’s assistant leader was killed, Soutra repeatedly stepped into the line of fire as he directed the rest of his team to fight off the enemy. He also dragged several wounded men to safety during the ordeal.

During the ceremony, secretary of the navy Ray Mabus said, “This is a chance to recognize people who don’t get recognized much.” He added that the actions of the entire team reveal “just how incredibly capable MARSOC operators are.” Such capability, coupled with a gung-ho willingness to lay their lives on the line to save their team members, ensures that MARSOC will remain one of the key elements in the worldwide War on Terror for as long as they are needed.

Helmeted MARSOC dogs on patrol assist their teams in escorting foreign politicians.

FORCE FACTS MARSOC includes some K-9 units, which consist of dogs that are specially trained to sniff out explosives.

Marines are trained to keep all senses alert in every environment.

FORCE FACTS MARSOC works closely with Navy SEALs and Army Special Forces (the Green Berets) to make sure that their battlefield skills complement one another's.

GLOSSARY

amphibious operations – assaults from the sea that involve coordination of naval, airborne, and infantry components; they are generally regarded as the most complex and potentially riskiest of all military maneuvers

coordinates – where the longitude and the latitude of a point on Earth intersect

enlisted men – those who sign up voluntarily for military duty at a rank below an officer; they compose the largest part of military units

fast-roping – sliding down a thick rope suspended from a helicopter as rapidly as possible

forward operating bases – bases established in friendly territory near enemy lines to provide support for tactical operations

Global Positioning System – a navigation system using 24 orbiting satellites located more than 10,000 miles above Earth

Hells Angels – members of a famous American motorcycle gang noted for their fearsome appearance

humanitarian aid – food and other supplies delivered to people who have been adversely affected by wartime conditions

improvised – produced without advance preparation from available materials

infiltrations – secretive passages into enemy-held territory

insurgency – an unorganized rebellion against a government

mercenaries – people who are paid to take part in armed conflicts

prismatic scope – a telescope with the sighting device mounted on the top or on the side

reconnaissance – a search to gain information, usually conducted in secret

squad automatic weapon – a lightweight machine gun offering a portable source of automatic weaponry to a small unit

steroids – manmade chemical compounds that contribute to increasing the size and strength of an individual

Taliban – a fundamentalist Islamic political movement and militia that controlled Afghanistan; noted especially for terror tactics and a repressive attitude toward women

trauma – extreme shock to the body resulting from stress or injury

wadi – a narrow, steep-sided ravine or gully in southwestern Asia and northern Africa

FORCE FACTS For swift entry into an enclosed space, operators may use a stun grenade that produces a blinding flash and ear-shattering explosion, disorienting potential enemies without killing them.

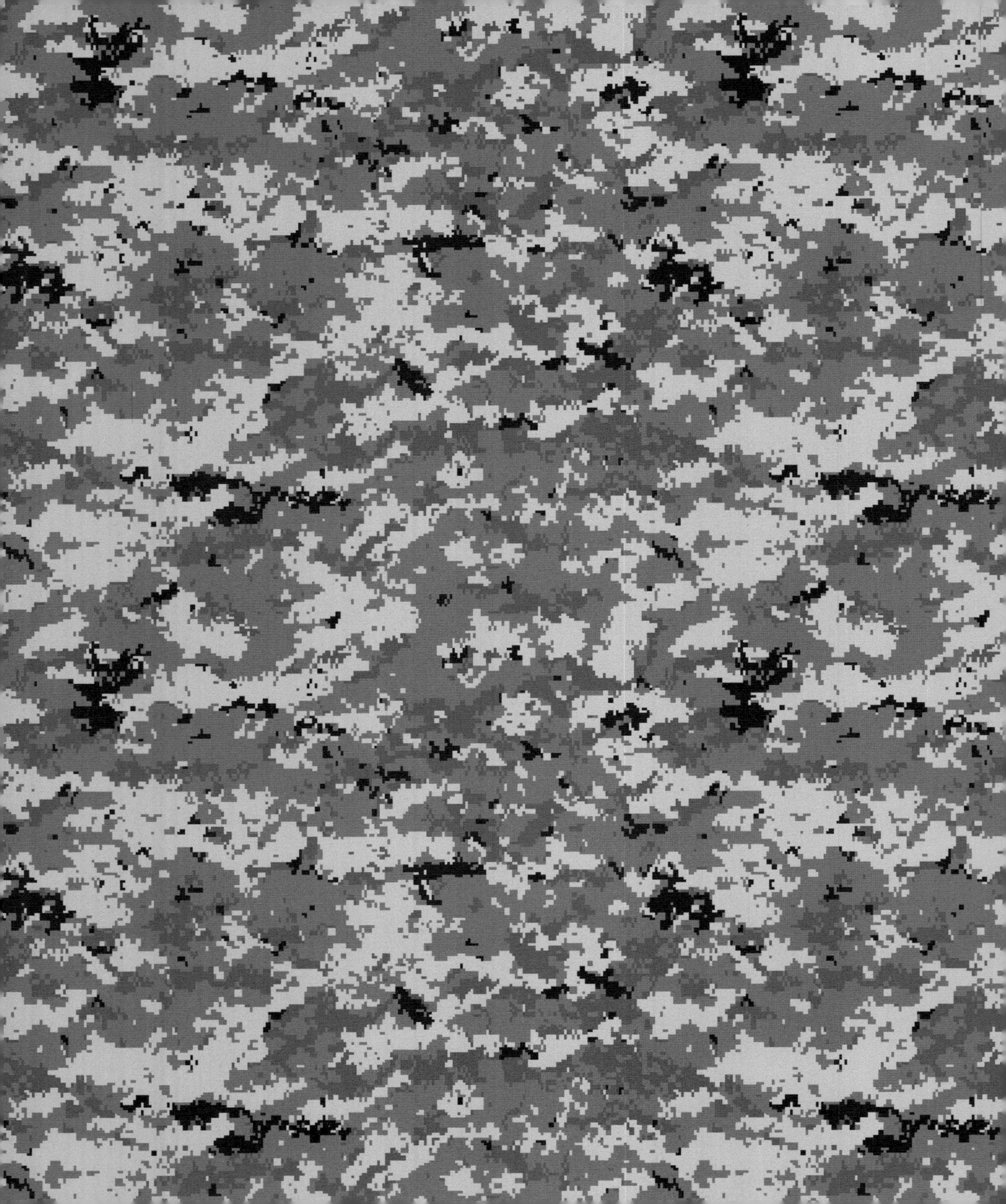

SELECTED BIBLIOGRAPHY

Cantrell, Mark, and Donald Vaughan. *Special Forces: America's Elite*. Bonita Springs, Fla.: The Media Source, 2012.

Chant, Chris. *Special Forces*. Bath, UK: Parragon Press, 2012.

Frederick, Jim. *Special Ops: The Hidden World of America's Toughest Warriors*. New York: Time Books, 2011.

North, Oliver. *American Heroes in Special Operations*. Nashville: Fidelis Books, 2010.

Pushies, Fred. *Marine Force Recon*. Minneapolis: Zenith Press, 2003.

———. *MARSOC: U.S. Marine Corps Special Operations Command*. Minneapolis: Zenith Press, 2011.

———. *Special Ops: America's Elite Forces in 21st Century Combat*. St. Paul, Minn.: MBI Publishing, 2003.

Tucker, David, and Christopher J. Lamb. *United States Special Operations Forces*. New York: Columbia University Press, 2007.

WEBSITES

SOFREP: About MARSOC

http://sofrep.com/marsoc/

This Special Operations Forces Situation Report focuses on MARSOC's history, weapons, and a "typical day."

U.S. Marine Corps Forces – Special Operations Command

http://www.marsoc.marines.mil/

The official MARSOC website includes news, photos, historical details, job descriptions, information about selection and training, and more.

READ MORE

Brush, Jim. *Special Forces*. Mankato, Minn.: Sea-to-Sea, 2012.

Cooper, Jason. *U.S. Special Operations*. Vero Beach, Fla.: Rourke, 2004.

Note: Every effort has been made to ensure that the websites listed above have educational value and that they contain no inappropriate material. However, because of the nature of the Internet, it is impossible to guarantee that these sites will remain active indefinitely or that their contents will not be altered.

INDEX

Afghanistan 9, 12, 27–28, 35, 36, 38

Africa 11, 15

Carlson, Evans 38

cooperation with other special forces 14, 43

deployment 17, 20, 27, 33

early MARSOC history 12, 14–15, 30, 31, 38
- birth of organization 14–15
- Detachment 1 14
- Force Recon 12, 14, 30, 31
- Marine Raiders 12, 38

Garrity, Patrick 32

headquarters 16, 23
- Camp Lejeune, North Carolina 16, 23
- San Diego, California 16

Hejlik, Dennis 15, 19

Iraq 12, 14

Lefebvre, Paul 40

Marine Corps hymn 11

media portrayals 30–32, 35
- films 30–32

military decorations 12, 36, 39, 40

missions 15, 16, 19, 24, 27–28, 33, 35–36, 38–39
- and forward operating bases 36, 38
- in Panerak 38–39
- Red Thunder 35–36
- types of 27–28

Mosser, John S. 36

motto 12, 38

nicknames 9, 16

organizational structure 15–16, 20, 24, 27
- battalions 15, 16, 20, 24, 27
- companies 15, 16
- regiments 16

personnel 9, 14, 15–16, 19, 20, 24, 25, 27, 30, 36, 40, 41
- K-9 units 41
- support groups 16
- teams 15, 16, 20, 30, 36, 40
- *See also* Special Operations Teams

qualification tests 20

reputation 35

responsibilities 24, 27–28, 36, 38
- direct action 27
- foreign internal defense 27
- special reconnaissance 27
- village stability operations 27–28

Rumsfeld, Donald 14

Soutra, William, Jr. 40

Special Operations Teams (MSOTs) 15–16, 17, 20, 24, 30
- critical skills operators 15, 16, 17, 20, 24
- joint terminal attack controllers 16

Taliban 9, 35, 36, 38

terrorism 12, 14, 15
- September 11 attacks 14

training 7, 17, 20, 22–24, 28
- close quarters combat 22 , 24, 28
- Derna Bridge exercise 24
- fast-roping 20, 24
- field skills 22
- firearms 23–24
- individual training course 20, 22–24
- irregular warfare 23
- languages 20, 24
- parachuting 7, 24
- physical fitness 20, 22
- rappelling 24
- small-unit tactics 22
- Survival, Evasion, Resistance, and Escape 22
- swimming 22, 24
- Tactical Combat Casualty Care 22

transportation 9, 16, 17, 20, 22, 24, 38
- aircraft 20, 24, 38
- animals 17
- boats 22
- dirt bikes 9
- Ground Mobility Vehicles 16

U.S. Special Operations Command 14, 20

War on Terror 32, 40

weaponry 9, 16, 22, 28, 30, 45

Woitowicz, William 38, 39